Sahara Desert

Exploring the World's Greatest Desert on Foot

(The History and Legacy of the World's Greatest Desert)

Jensen Ullrich

Published By **Zoe Lawson**

Jensen Ullrich

All Rights Reserved

Sahara Desert: Exploring the World's Greatest Desert on Foot (The History and Legacy of the World's Greatest Desert)

ISBN 978-1-77485-932-2

No part of this guidebook shall be reproduced in any form without permission in writing from the publisher except in the case of brief quotations embodied in critical articles or reviews.

Legal & Disclaimer

The information contained in this ebook is not designed to replace or take the place of any form of medicine or professional medical advice. The information in this ebook has been provided for educational & entertainment purposes only.

The information contained in this book has been compiled from sources deemed reliable, and it is accurate to the best of the Author's knowledge; however, the Author cannot guarantee its accuracy and validity and cannot be held liable for any errors or omissions. Changes are periodically made to this book. You must consult your doctor or get professional medical advice before using any of the suggested remedies, techniques, or information in this book.

Upon using the information contained in this book, you agree to hold harmless the Author from and against any damages, costs, and expenses, including any legal fees potentially resulting from the application of any of the information provided by this guide. This disclaimer applies to any damages or injury caused by the use and application, whether directly or indirectly, of any advice or information presented, whether for breach of contract, tort, negligence, personal injury, criminal intent, or under any other cause of action.

You agree to accept all risks of using the information presented inside this book. You need to consult a professional medical practitioner in order to ensure you are both able and healthy enough to participate in this program.

Table of contents

Introduction _____ 1

Chapter 1: The Ancient Sahara _____ 5

Chapter 2: The Trans-Saharan Trade ___ 40

Chapter 3: Saharan Slaves _____ 73

Chapter 4: The Modern Sahara _____ 100

Chapter 5: Water _____ 107

Chapter 6: People And Cultures _____ 115

Chapter 7: What To Do Within The Sahara Desert _____ 127

Chapter 8: Antarctica _____ 136

Chapter 9: "Mirages": One Of The Greatest Desert Mysteries _____ 150

Conclusion _____ 154

Introduction

The image of Luca Galluzi's desert in the western region of Libya

The Sahara

It is said that one is exposed to a shocking glimpse of what is best defined as an existence crisis walking through the golden and infinity-spanning landscape in the Sahara Desert. If one went up any of the many dunes of sand and then twirled in a particular area, especially in the night it would be difficult to in a position to distinguish the east and west, without using a compass since the landscape would consist of nothing more than stunning carpets of indistinguishable sandy hills that loomed over the vast horizon.

Astonishing thoughts of goal and self-worth aside, those lucky enough to get to this beautiful sandscape with a big SUV or, more commonly, a camel caravan that is hired is presented with a stunning view. The waves that lap the stunning assortment of irregularly-shaped dunes of sand, some only half-a-story high, and some up to 600 feet in the height, are as distinctive as the marks on your fingertips. Some say they're the stretch marks of Mother Nature. For others, the wind-driven patterns could conjure the image of a god holding a rake while whirling haphazardly and sifting through the grity particles of sand, just similar to the zen-like garden of a miniature to take away the stress.

It's not difficult to comprehend the reason why some people feel inadequate and insignificant, especially when engulfed in such a gorgeous, yet dangerous and almost unimaginably vast terrain. The scattered wildlife and flora squirmingly scattered throughout the huge expanse of barren ground will not be enough to soothe the stomach that is rumbling. Additionally, a slightly threatening monotonous sound pierces the silence created through the vibrating particles of sand that are crashing across the dunes. The musicians behind these monotonous music are none other than the pockets of wind that are massaging the sand dunes, and the unique phenomenon which evokes a chorus of conch shells crying or a buzzing hive of glowing bees, causes an electric tingle through one's spine.

Naturally, these fascinating sounds are nothing but boring to people who reside or travel through these areas often. Certain people have become accustomed the rhythm of the dunes so that they can't bear when they are not. The music it plays during the otherwise strenuous journeys they take allows the self-reflection and contemplation reflections. Many feel a sense peace by listening to the soothing harmonies of the dunes' crooning. However, for a handful of people those who listen, the sounds of the dunes are more than just white noise. Take a close look to them, they say that you could hear the sand sing about the captivating historical events and the fascinatingly complicated life that define this magical desert.

Chapter 1: The Ancient Sahara

"I have always been a fan of the desert. Sitting on a desert sand dunes and is unable to see, hear or even feel nothing. In the silence, however, something is pulsing and shimmers ..." - - Antoine de Saint-Exupery. The Little Prince

The Sahara is as deadly as it is esoteric and mysterious, is the source and the setting for a myriad of myths. These tales that have been passed down from generation to generation over the course of centuries, usually focus on the lost oases and hidden gems from the Sahara The most famous of these abandoned paradisiacal cities being the mythical Zerzura.

Zerzura was first mentioned in a treasure hunt pamphlet called the "Kitab al Kdnuz (Book of Hidden Pearls) which was written by an unknown author around 1400. According to this unknown fortune-seeker, Zerzura was once upon an era a magnificent city, awashed in white sand, and stunning man-made structures that were in various shades of ivory, surrounded by four walls, and a huge gate, that was topped with a stone carving of birds. Because Zerzura was home to many of the world's most precious artifacts and treasures, it was hidden away in one of the less-visited areas of the Western Sahara, just "2 or 3 days due west from Dakhla." According to the legend states only those who were able to locate the hidden oasis by following these vague directions deserved the privilege of stepping foot in the sanctuaries.

Locating the exact coordinates of Zerzura However, it was not the only step. Only after overcoming the fierce Djinn and an army of pacing "black giants" that were stationed at the gate, could anyone be allowed to enter the city. Obsessed with the riches hidden at the gates only a handful of adventurers from outside briskly charged out, and were immediately struck by the Djinn's spears. The most intelligent of invaders, on the other on the other hand, either had sufficient reinforcement or were able to come up with a strategy that allowed them to sneak over, and ideally, sleeping guards. After a clean coast was granted to enter the city, they were charged with shifting the gate open and then snatching a unique key from the bird's beak , allowing them inside the city. Inside, the brave souls then had to walk through the castle, in which they would see the queen and the king within the cobweb-strewn and dust-coated royal bedroom and their "enchanted" couple ensconced in a secure and deep sleeping. In the bed next to them were glittering gold heaps jewels, precious stones, and unique treasures.

Many adventurers packed their bags and their clothes with as much valuables as they could transport back home, doing it with the greatest care and deceit they could muster, in case the royals suddenly awake from their sleep. One knight, unhappy with the haul and woke up the charming aged queen with a sweet kiss, then proceeded include her in his riches.

It is not known much about Zerzura beyond the fact that it eventually fell into the depths. Some of its treasure trove of treasures because of these treasure hunters and their treasure hunters, are scattered throughout the Sahara. Much like the cityitself, was never returned to.

While we can't deny the value of entertainment these stories of Zerzura offered the tale was dismissed as nothing more than sparks for a campfire were it not due to the story of Hamid Keila, recorded by the Emir's scribes from Benghazi. It was 1481 when Keila had been traveling in a caravan heading to the oasis towns in Dakhla and Kharga was confronted by the most severe sandstorm that he ever seen in his life. The sandstorm was so dramatic in its fury that it continued to rage for more than one week. At the point when the dust had finally was blown away all the caravans with the exception of Keila was suffocated to death.

For several weeks, he walked across the Sahara on his own and was severely malnourished and dehydrated and as an aforementioned, borderline delusional man was found by a solitary group of tourists. These slim and tall men, with beautiful blue eyes, light hair that was as gold as the sand under his feet as well as "straight swords" instead of scimitars were so alien to me that it required some convincing from them for him to drop his guard. He was unaware that the friendly strangers had lifted Keila to an animal and were taking the man towards their city eager to help him return to the health he was in.

It was only when Keila had been served lots of food and drink that Keila could be able to appreciate the amazing landscape, which he soon discovered was

Zerzura. He had never before seen an area with such vibrant and diverse wildlife. It was further adorned with the sound of bubbling springs, lush grasslands, stunning lakes and other unusual plants. Keila was particularly enthralled by the city's incredibly advanced multi-story structures that were "white like a dove" and the small flocks of starlings who set up their camp among the seemingly infinite tree. It is interesting to note that the residents that were cared for by Keila who was the city's chief, were referred to as "El Suri," and spoke a dialect of Arabic that he did not know. Additionally El Suri were El Suri were believed to be non-Muslims, due to the absence of mosques, and the naked women.

It was the Emir of Benghazi initially welcomed the previously dislocated Keila to his town with open arms however, he began to become skeptical of the shrewd traveler and finally called him to his residence. After a heated exchange in which the Emir assessed as well as filled in numerous gaps in the traveler's account, Keila was revealed to be a wanted man. He was a victim of the generosity of his rescuers, and then repaid them by taking a portion of the riches from the castle of the royal family. The Emir immediately ordered his guards to take a pat Keila down. Among the silver, gold, and sparkling trinkets he removed from his body was a stunning ruby encased in an elegant band of pure gold. Keila maintained that the items in his possession were family treasures, but not just did the Emir disbelieved, but he also confiscated the loot, but he also snatched it off the hands of Keila and banished Keila from the city.

The ring of rubies, along with the other precious objects Keila brought from the secluded city of Zerzura was eventually placed in the hands of Idris I Idris, the Emir of Cyrenaica and the first King of Libya in the early 20th century. The experts employed by Idris identified the ring's hallmarks from the 12th century European workmanship, leading many to believe that it was once owned by "Teutonic Arabs" the name used to describe the crusader knights who had dissociated themselves from their fellow knights in their journey to The Holy Land, and though stuck in the Sahara survived in the scorching desert heat for a number of years afterward.

King Idris I

Apart from the Ruby ring The ruby ring aside, nobody is sure what the rest of the Zerzurian treasure contained. Many believe that it was the case that El Suri hoarded ancient Egyptian gold, bejeweled chalices and other precious relics they had stolen from the Pharaohs. Some speculated on the existence of mysterious manuscripts or maps that led towards the Fountain of Youth and other mysterious scrolls that contained information about the esoteric and spiritual.

Although there is a glaring lack of proof to suggest that this treasure or even the city, or the city itself, existed, a lot of people of us, from mavens and scholars to treasure hunter, remain fascinated by the mystery. Along with engaging in numerous debates regarding the presence and location of Zerzura that is documented by their journals they also Royal Geographical Society sponsored multiple explorations into the Sahara. They were so determined to find the treasure that they used airplanes specifically for this purpose in the very first instance. Sir Robert A. Clayton of the Society observed, "Though aeroplanes had not before been used to explore in this area of the world, we believed our aircraft could aid us by expanding our range of our vision and allow us to study mountaintops ..."

Is there something about Sahara that inspires such compelling legends? To gain an understanding of the mystery surrounding the desert it is necessary to first go over the fundamentals. To begin it's true that Sahara is the Sahara is the biggest "hot desert" currently in existence, yet it is only ranked 3rd in the overall ranking of deserts, lagging just behind the cold Arctic and Antarctic Polar deserts. At around 9.4 million square km (roughly 3.6 million sq miles) The Sahara is nearly the size in North America, makes up 10 percent of African continent. The Sahara, Arabic for "The Great Desert," covers the majority the region of North Africa, sweeping across vast areas in Algeria, Chad, Egypt, Libya, Mali, Morocco, Niger, Sudan and Tunisia as well as other. It is surrounded by the Atlantic Ocean, as well as the Mediterranean and Red Seas, to its west east, north, and west and north, respectively and east, the Sahara is further divided into various sections that include The Tibesti Mountains; the Air Mountains and the Central Hoggar Mountains; Western Sahara as well as The Tenere as well as the Libyan Deserts (the latter the dryest within the area) as well as a variety of other areas of desert plateaus and mountain ranges.

Although the Sahara is often portrayed as flaxen flatlands surrounded by sandy hills as well as desert mountain ranges, the vast majority area of deserts is known locally as"a "hamada," its topography is characterized by rocky plateaus as well as rough, barren lands that are coated with sand. Ergs, or sandydeserts with dune-filled dunes make up only 25% of Sahara.

A picture of Saharan dunes

The less than an inch rain falls on half of the Saharan land each year, and the rest is humidized by 4 inches. It is no surprise that the frequent dust devils and sandstorms, which are the result of Sahara's distinctive "northeasterly winds" as well as the absence of rain, can all be blamed for the lack of humidity as well as the scorching desert heat. It is that extreme that these brief bouts of rain, even though they are often described as "torrential downpours" are rarely perceived by people within the Sahara because the showers disappear in a matter of minutes after being sparked by the clouds, with drops of rain falling away before they ever touch the ground. The disappearing rain, signaled through "streaks" in the mist under the glistening clouds, is referred to as"virga. "virga."

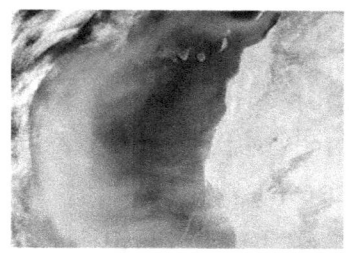

A satellite image of a sandstorm originating from the Sahara moving across the Atlantic

Understanding and studying the Sahara's constantly changing array of ecoregions has helped nomadic tribes and explorers over the centuries to identify the most suitable areas of the Sahara and to understand the special assets. A few of the important regions is the "Atlantic Coast Desert" that, as its name suggests, is the narrow strip of land which is bordered by Atlantic. With an area of around 15,400 square miles The desert's coastal stretches across the southern part and west of Morocco as well as Mauritania. It is regularly cooling with the Canary Current drifting from the sparkling waters, this area is one of the coldest of Sahara's ecosystems, with a temperature range of 27oC (80.6oF) as well as periodic drop up to 13oC (55.4oF). The fog that is a result of the Canary Current provide much-needed relief to this humid stretch of land, it stimulates the development of drought-resistant trees succulents, euphorbia, succulents and large swarms if the lichen.

Rossella Piccinno's photo of a portion of the Atlantic Coastal Desert

The region, which is known as"the "North Saharan Steppe and Woodlands," about 646,000 square miles located on the northern reaches of Sahara. Surrounded by its famous dunes, it is home to numerous physical characteristics, from the geologic marks and rocks, to ravines and wadis. In spite of the fact that this ecoregion has the most virgas, the four inches of precipitation that is received by the area each year appears to be adequate for different species of snakes that inhabit the area scorpions, lizards and scorpions birds, vultures as well as some mountain

gazelles.

Jim F. Bleak's photograph of Souss-Massa National Park located within the North Saharan Steppe and Woodlands

In this ecoregion, guests will discover an area known as the "South Saharan Steppe and Woodlands," which covers part of the Western Sahara, about 425,400 square miles of surface area. The availability of water is very limited in this deserted area, especially during summertime. This being said the land that is prone to drought that extends from Mauritania from Mauritania to Sudan is not low on xeric trees and dry forests. It includes wildlife, herbs and other species that are unique for the area.

Next is what geographers call"the "West Saharan Montane Xeric Woodlands." This section of land, approximately 99,700 square miles encompasses a significant portion in the Saharan highlands Stone plateaus, stone hills, and plains of gravel which are primarily made up from volcanic lava. As with the rest of Sahara region, this one is shattered and drained by the scorching heat of the long summers, however the cool, rainy winters, along with its elevation, have made ideal for blooming of thorn and athel pine trees. Swifts, warblers, larks as well as other birds have a tendency to stop their migration to rest for a few minutes in the woodlands of western Africa.

Bertrand Devouard's photo of Ahaggar National Park in the West Saharan Montane Xeric Woodlands

In the central region of Sahara there is located the "Tibesti-Jebel Uweinat MontaneXeric Forests" which encompass two highland regions. It covers about 31,700 square miles It is located primarily in Chad which is a thin portion of the forest spilling into Libya. Contrary to the ecoregions of its neighbors The area is blessed with a higher quantity of rain. This allows for the gradual growth of lush grasslands, as well as other types of life found in the Mediterranean like lavender, figs, as well as olive tree. Warm-blooded animals, like the sand-burrowing jerboas and gerbils and other rodents, aswell like antelopes, cheetahs, and screw-horns (the latter being the biggest native mammal of the Sahara) as well as antelopes and cheetahs can be found in these regions.

Michael Baranovsky's image of a part of Tibesti Jebel's Uweinat Montane Xeric Woodlands

"Saharan Halophytics. "Saharan Halophytics" are found throughout the Sahara located in areas that are closest to saltwater deposits like salt lakes like the Tunisian salt lakes and which include the Chott Melghir in Algeria, the Siwa and Qattara depressions of saline in Egypt and other smaller areas of land in regions like the Western Sahara and Mauritania. They are also found within Algeria, Mali, and Niger in addition to the western portion in the Hoggar Mountains, is an ecoregion dubbed"Tanezrouft. "Tanezrouft," said to be among the most hot and driest regions in the Sahara.

The last, but certainly not last is the geographers' term for"the "Sahara desert" ecoregion. Like those of the North Saharan Steppe and Woodlands The 1,783,000 square miles of desert that lie in central Sahara is made up consisting of dunes, sand terrains flatlands that are saline and massive, snow-capped mountains. In spite of its diversity it is afflicted by dusty, savage temperatures and unforgivingly dry winds and its temperatures, at times, climbing to 40-55oC (104-131oF). The species of fauna and flora here have gotten forcibly resistant to temperatures and humidity.

Evidently, this diverse terrain didn't appear in a flash, so how did the Sahara develop into the landscape it is currently? Religious and superstitious people look to the ancient creation myths to provide a deeper understanding of the Sahara's beginnings. There are those who believe that Allah was searching for an enclave of sacred meditation, stripped the Sahara of vegetation to allow him to wander in peace and unaffected while they sorted and collected their thoughts. Some believe the reason was Neter Seth who was one of the Egyptian Masters of Disasters and the so-called "Lord of the Sahara," who transformed the Green Sahara into the Red Desert by the unbridling in the Great Deluge. It is believed that the Great Deluge, also known as the "Great Flood," was Seth's answer to rid the world of debauchery, corruption and other sins of the mortal realm.

While these mythological stories of origin could be, historians and geologists use science to draw an even more accurate image of what happened to the Sahara really developed. For many centuries it was believed that the Sahara was believed to be between 2 and three millions years old. In August 2014, scientists from the Bjerknes Centre for Climate Research (BCR) located within

Bergen, Norway, tested samples from the sand dunes in the Chad Basin and concluded that the samples were at the very least 7 million years old. Models developed by this institute proved that, 7 millennia ago the average amount of rainfall within the area was nearly twice that amount.

After greenhouse gas, the periodic changes in the Earth's tilt and the development of vegetation were excluded by scientists as the cause of the declining moisture levels in the Sahara and they were left with just one possible explanations: tectonic changes.

About 250 million years ago an immense area of water, referred to in the "Tethys Sea" was encased by two supercontinents, Laurasia as well as Gondwana. As these land masses fell apart and the slabs of land began to move into the ocean then it was the time that the African as well as the Eurasian plates collided into each other. The collision did not just trigger the creation of the Alps and the Himalayas The kissing edges of these plates also functioned as gates that sealed off some that was part of Tethys and spawned the Mediterranean Sea.

The dehydration of the once green continental area of Africa began with the time that the western arm of the Tethys was replaced by the area now known by"the "Arabian peninsula." The huge piece of land that is now connected to the coast deprived the continent of most of its sunlight, which in turn changed the patterns of weather in the region. It was the African continent continued to cook in the sun as the years passed by and the temperature climbed upwards each 40 years, dictated by the rising angle of Earth's axis. This was the reason that vast areas of vibrant forests and land started to die off.

The landscape change is so drastic that Sahelanthropus tchadensis species, which is among the oldest species of the family tree of humans that had to this point lived in the northern reaches of Chad close to the Sahara was left with no option but to invent a solution to avoid becoming extinct. While a small percentage decided to remain in the area and the rest migrated to the south in search of more moist and more fertile areas. The researchers at the Bjerknes Center concluded that the changes in precipitation played a key role in this "evolution as well as dispersal of the hominins from the north of Africa."

It turned out that the changes only began. Over the next six million years or more, Mother Nature continued to change Africa by the triggering of more earthquakes, and the triggering of several Ice Ages, which brought about the shifting of dry and wet conditions across the Sahara. The Sahara had the wettest years and the thawing of each Ice Age, until the final glacial period ended around 11700 years ago. It was a time when the continent Africa gradually defrosted and in the course of the Holocene Optimum that occurred around 9,000 years ago, the temperature and rainfall reached an all-time high. The people living in the region were thrilled by the rapid increase during monsoons (which were once a source of water for the western part of Africa as well as the Mediterranean) as well as the emergence of new plants and foliage within what scientists have called"the "Green Sahara." About 3000 years before the time of this writing the only viable regions of land were the eastern part of the Sahara particularly that of the Nile Valley. Different tribes battled one another to claim a place in the valley, leading to bloody battles and even fights. A minimum of 45 percent of the dead of those buried in Jebel Sahaba cemetery Jebel Sahaba Cemetery belonged to tribesmen who were killed fighting for a share of the area.

In the end, the tribes left the once-congested Nile Valley and spread out across the entire region and eager to make use of the newly-discovered resources. Many moved to the highlands and hid in the shade of the looming cedar and oak trees. Others chose to set up their camp in the lowlands, that had been transformed into an "savanna grassland" that was now covered with acacia and hackberry trees, along with flowing rivers and sparkling lakes of saltwater and freshwater lakes, such as Lake Mega-Chad, then the largest freshwater lake in the world as well as the Wadi Howar, which was the biggest river that fed the Nile from the Sahara in the period of.

The tribesmen weren't the only ones to have migrated into the Green Sahara. Paintings that are faded but that archaeologists have preserved in the relevant caves indicate that the humans were joined by animals that were of every kind. Elephants, lions, giraffes rhinoceroses, ostriches as well as other mammals of the land swarmed and ate the grasslands. Crocodiles, frogs, hippos and more than thirty species of fish including catfish and tilapia wading in the water.

The people who settled in this area were later to be recognized as "Lost Tribes from the Green Sahara." As the majority of these tribes failed to record any written documents, experts have only theories regarding their history. The majority of them bowed before matriarchal goddesses who represented the moon and sun and chose the "Moon Calendar" as their official calendar. People living in Nile Valley Nile Valley, on the contrary, worshipped Nuth which was also called The Holy Spirit of Amon which was the Egyptian goddess who presided on the Lake Tritonis in Nubia, and the surrounding islands.

The Saharan tribes lived alongside animals in harmony. Hunting was not a pastime but was a necessity. Gatherers gathered grains and fruits during the day, but meat was essential to eat properly. Thus, hunters Armed with spears and bows hunted in the forest but before they killed their prey they sought permission and forgiveness of Mother Nature before doing so.

It wasn't until the start in the 20th century when scientists discovered evidence that helped them gain a deeper insight into these tribes that have disappeared. In the month of October, 2000 an paleontologist by the title of Dr. Paul Sereno from the University of Chicago led his team to the Niger's Tenere Desert. After removing the fossils of a huge crocodilian reptile, as well as an immense dinosaur that had 500 teeth that were jagged, Senero was keen to keep up his impressive streak. After arriving the team assembled, pulled out their toolsand got to work in a flash. The workers worked on the earth for several hours However, as the hours passed, their enthusiasm started to diminish. Mike Hettwer, the photographer Senero had contracted, became uneasy and wandered away from the site of excavation. It was on his leisurely walk that he stumbled upon three dunes that were "literally covered in bone fragments." Senero and the crew rushed to the exuberant Hettwer immediately to examine his discoveries. While it became apparent that the fossils were not related to dinosaurs but rather humans The crew was thrilled as they discovered bones belonging to that lost Green Sahara tribe.

Within the many Human remains, were evidence about the life the tribes lived including "clay beads, potsherds and stone tools" as well as a variety of animal remains. These findings inspired Sereno to transform his research. After putting his passion for paleontology off, he returned to the region repeatedly throughout the next few decades, determined gather the most information possible about the tribesmen who have disappeared. In the year 2003 the researcher discovered seventeen burial places.

Despite the significant advancements he was making there was a gap in his knowledge of this particular field. Therefore, he sought out the assistance of renowned archaeologist Elena Garcea. They returned to the site of excavation that is now called"the "Gobero," the oldest graveyard in the Sahara and later in the year. After a few days the excavation, they found another pottery fragment that was decorated with a distinct "pointillistic style" and linked it to the nomadic tribe of fishermen who used to roam these areas, known as"the "Tenerians." A few days after, she demolished a different ceramic shard. Observing its "wavy lines" drawn on the surface, she linked it to the Kiffians who were a hunter-gatherer clan which was founded around one millennium after.

The Tenerians as indicated by their name resided at the Tenere Desert, a hot and savage slice of the Sahara which the locals call the "desert within an oasis." As evidenced by the size and the density of their bones and teeth they were Tenerians were slimmer and less bulky than Kiffians. Due to fishing equipment and other hunting equipment that surpassed their peers and made them live more relaxed and less stressful lives.

It's possible that they were one of the most superstitious; the bones and tusks of crocodiles warthogs, hippopotamuses and other animals were intentionally placed with the remains of many corpses that gave archaeologists an insight into the customs of the Tenerians' celebrations. Pollen residue that was found beneath certain corpses which included the mother and her infant, revealed that the bodies of some were laid in a flower-filled bed.

Since the Tenerians consumed the majority of the fish found in the Sahara's lakes that were shrinking and rivers, the Kiffians as well as the younger generations of the tribeswere pressured hunt animals on land for food. Evolution seems to have prepared them for this job, since they were strong-built. Both genders had naturally huge and prominent muscles in the arm and leg and were as quick as they were powerful. Some were as tall as 6 feet. Yet, the lack of wear and wear on their bones as well as the rareness of "head as well as forearm injuries" revealed to researchers that they were "peaceful and hard-working group" who generally stayed away from the raging territorial conflict in the time. Due to archaeologists' inability of finding certain Kiffian artifacts within the Gobero area, very little is known about the hunters-gatherers, aside from their use of harpoons as well as their high protein diets.

The discovery of the Gobero helped researchers improve the understanding of and the Tenerians, Kiffians, and one of the unknown Libyan tribes who lived in the Uan-Muhuggiag region located just 1500 miles to the west of the Nile Valley. In 1958, professor Fabirizio Mori from the University of Rome traveled to the southwest of Libya in the southwest of Libya, where he stumbled to a cave, and discovered a remarkable discovery that would be the most memorable moment of his career. It was the Tashwinat Mummy, or the "Black Mummy of Libya," which is thought to be 5,400 and 7,700 years old. It predates the Egyptian Mummies by at least two thousand years. After surgery to remove his organs the child that was no more than 3years old, was given an organic preservative, and then manipulated in such a way that he was rolled into the position of a fetal. The body is then dismembered, shaved down with different herbs (presumably to cover up the smell and the slow process of decomposition) and then wrapped in skin of antelope or goat. A necklace made of eggshells of ostrich was also discovered attached to his neck.

When they examined drawings and cave paintings archaeologists discovered that the people who created the body were cattle farmers who also resided in the Sahara during the Savanna era. The other artifacts found in the cave, like the plethora of stone tools as well as an "horned animal skull" that was used in the role of the "emblem of the sun" were left by tribespeople from different eras.

Chapter 2: The Trans-Saharan Trade

"Aut throughm inveniam faciam." ("I I will find a way to do it to make one, or find the one"). It is believed to be the work of Hannibal of Carthage

The transition from desert to savanna is marked by two periods. The first was between 4700 and 3500 BCE and the second took place in between the years 2000-1700 BCE. After the monsoons moved to the south, and the winds came back to Mediterranean and the remaining remnants of the declining grasslands filled with bald spots as well as shallow lakes, quickly disappeared. In 2000 BCE in the year 2000 BCE, the Sahara was just as trocken as today. The final of the lakes evaporated around 1000 years later, several members of Saharan tribal groups, following the steps of their forefathers and migrated north south, east, and west, to find more lush grasslands.

Steph Lulu's image of a mountain created by wind

People who came to northern Africa are thought to be the ancestors from people like the Egyptians, Phoenicians, Greeks, Jews, Berbers, and Tuaregs along with Nubian (Cush), Ethiopian (Aksum) as well as Sudanese (Ta-Neti) groups. The tribes that developed here were able to speak in dialects and languages that were similar to that of Ancient Egyptian tongue. People who traveled south gave birth to those who were Dravidians in

South India and the Bantus of South Africa, as well as the ancestral Pakistani as well as Iranian tribes. The ones who traveled east were the forefathers to those who came from the Yoruba, Mande, and the Dogon in West Africa. Others, who went west to settle in the area that would later be Babylonia.

The tribe who bravely chose to not seek their fortune elsewhere were the Garamantes who were a shrewd and extremely resourceful people who resided in the 250,000 square miles of the present-day Fazzan province in southwest Libya. As with the other tribal groups of the Green Sahara, virtually nothing was known about the Garamantes for quite a while The historians only had the narratives of biased chroniclers like in the fifth century CE Greek writer Herodotus as well as various other Roman writers. Herodotus described them as uncultured farmers with a simple mind who covered their fields with salt using Humus (a organic dark substance made by a combination of decayed leaves and animal matter). A passage from Herotus's observations of the Garamantes was explained as follows "[The Garamantes] avoid all interactions with men, and have any weapons for war and are unable to defend themselves. ...[Theyare hunted by for the Ethiopian hole-men, also known as trogloodytes, riding in 4-horse chariots because these troglodytes are extremely swift on foot...They consume lizards and snakes and reptiles and speak in a dialect like none other, yet they have a squeak that resembles bats. ..."

A spherical bust from the time of Herodotus

The Romans are believed to have engaged in war with the Garamantes in a variety of occasions and, therefore, not surprising, the stories they penned about the Garamantes were less than flattering. Tacitus who was one of the most famous Roman historian of the 1st century of CE, condemned the Garamantes in the form of "a wild tribe prone to plundering...nomads who moved from inaccessible camp to another." Some portrayed them as a vile, subhuman brute.

David Mattingly, a professor from Leicester has shed a brighter light on the Romans attitudes towards Garamantes: "Consider the epithets they use in their description: many and fierce, savage, indomitable...naked, miserable, dwelling in tents or huts and swarming, sexually promiscuous lawless, smugglers of booty...given to the brigandage trade, black." Black slaves were frequently referred to with a derogatory name "faex Garamantarum,"" as well as to use the term in English, "Garamantian muck." Mattingly adds, "The generally negative tone of these words must be understood as the nature of it as a mix of preconception and prejudgment."

The myths surrounding the Garamantes prevailed until a professor Charles Daniels from the University of Newcastle showed a distinct side to their story in the 60s. His research was further expanded by the mentioned David Mattingly. The revived enterprise, appropriately named the "Fazzan Project" was managed under the University of Leicester in conjunction with the Society for Libyan Studies in London and the Department of Antiquities in Libya between 1997-2001. The investigations allowed researchers from the British researchers to debunk the myths and exaggerations propagated by the Garamantes adversaries. It turns out that the Garamantes were in fact significantly more advanced than the enemies they fought to admit. In addition to establishing an amazingly intricate writing system They also practiced metallurgy, experimented with medical science, and controlled an impressive walled empire which flourished for a long time.

Garama is the main city city of the Garamantes, who first came into the spotlight during the early 4th century BCE. Garama was one of the vast Garamantian Empire. It was home to many other towns, more than 50 villages and some villages. There were a reported 4,000 residents in Garama during its peak and 56,000 Garamantes scattered throughout the 100-mile long Wadi Ajal (the depression the empire was buried in) as well as other "suburban village satellites."

In addition to their love of technological advancement In addition, the

Garamantes were also prolific architects. Their skills, which were unmatched, were seen in two structures, one of which was an impressive temple that was adorned with a colonnaded entrance and the short set of wide steps dedicated to Ammon, the Libyan god of the desert and the god for Jupiter, Ammon. A second one was a smaller structure that consisted of an impressive construction and a columned courtyard which was accessible only through a tiny entrance, which was enlivened by two "engaged pilasters."

The first structures built by the Garamantian masses were basic and box-like. They were built using the mud bricks. Mattingly's team swept away some of the mud-brick structures. The large rectangular buildings that were believed to be workshop or communal housing were made up of two to three rooms and included fireplaces inside and wells. In the Classic Garamantian period (1 - 400 CE) in the region, people began to construct grand stone structures made of ashlar for people in the upper classes of society. The commoners remained in their mud huts.

A collection of satellite and aerial photographs published in National Geographic allowed a fresh group of viewers from around the globe to take in the impressively well-preserved "Garamantian castles that lie in sandy areas." These Garamantian architecture, specifically its castles and fortifications were thought to be so revolutionary that the Romans copied their style in the later years, re-creating a strikingly similar fortresses throughout northern Africa and in the United States.

Artifacts discovered from the over sixty thousand Garamantian tombs found in Fazzan suggest an era of class-driven societies and a system that all the various and distinctive Garamantian tribes mostly adhered to. The majority of exhumed bodies were laid to forever rest in Cairns (memorial mounds of stone) or plain graves without markings. Others were buried in columns Romano-African mausoleums, as with mastaba "stepped" similar to pyramids constructed of mud bricks, and filled with 100 bodies per.

How did the Garamantes living in the middle of the Sahara and were able to avoid the luxury of such life? The first aspect of the answer is simple in control. It is important to understand that the Wadi Ajal was only the central point of Garamantian area; Garamantian chiefs controlled a large portion all of Trans-Saharan trade routes and also governed several additional wadis and oases that were beyond its borders. The chiefs were effectively doubled of their subjects, meaning that the roughly 120,000 inhabitants of the 64,000,000-acre (100,000 acres) area were expected to perform his will. The second aspect of the solution is in the Garamantes the healthy culture of trading which helped shape its booming economy. Traveling merchants carrying locally sourced salt gold, ivory, amazonite red carnelian, precious minerals as well as animals in cages (to be released into the arenas) domesticated animals and the natron (a type of alkali utilized for the preservation of corpses and in the manufacture of glass) went to Roman cities across the Mediterranean to market their products. The pottery, smithery and other items offered by salesmen on the move were supplied by a large system that included Garamantian workshops.

Apart from the money they earned from their exports, the bulk of the Garamantes' riches can be traced to their extensive involvement in the horrific but lucrative business of slave trade. To ensure that they could always replenish their slave stock, that could only be accomplished via war or conquest and conquest, the Garamantes had to be proficient combatants. They preferred to strike with their cavalry which consisted of a multitude of chariots drawn by horses. The Garamantian warriors wore a tinier armor, and protected themselves with nothing other than wooden round javelins and shields. However, they were agile, quick and agile. In addition, the daring warriors were just as brave as they were driven to power, racing into villages and towns in search of treasures and hostages in hordes. The captives were then made slaves and taken to the coast and auctioned off to the most affluent Roman bidders.

When you consider the thriving Garamantian economy, it's easy to comprehend why these people could afford materials and necessities easily however, how did they sustain their lives in such a harsh and deficient water source? It is evident that not being the people they were portrayed to be , according to the Romans They had to be creative. At the period of the Garamantes and they were able to see that the Saharan surface was filled with mineral deposits, most notably "white-crusted" calcium carbonate. Mineral deposits like these were remnants of crystallized lakes that were scorched dry by the relentless sun. The majority of the lakes were vaporized, however, there were sweet spots that contained secret storage chambers that contained "fossil water" enclosed within underground "permeable" rock formations sometimes referred to as "aquifers."

The aquifers would be their most valuable resource. In the past few generations the past, the Garamantes came across an intriguing aquifer at the bottom of the plateau, which was not far from the southern end of the Wadi Ajal. The researchers poked and poked at the wet, soft earth , and then inspected the area that was some feet higher than surface of the wadi. This was the moment they realized that they could redirect the underground spring to ensure that it would flow down into the valley, supplying access to water for all people living within the wadi. In a resemblance to Egyptian or Persian irrigation methods and practices, the Garamantes were able to drill through the ground by using double-headed hammers and iron gads creating conduits beneath the surface, referred to as "foggaras," breaching the aquifer , and transferring the water to different points in the valley. The channels were joined in their central oasis to create a inaccessible reservoir.

In all it was estimated that the Garamantes excavated 600 foggaras. This when placed side by side, would equal 3000 miles. The foggaras were protected by 100,000 evenly spaced "vertical shafts." The constricted conduits were identical in terms of height and width (5 two feet wide) however, they differed in length. Some served as connectors, while some were extended for miles.

The number of slaves and workers who perished because of the dangerous working environment either through overwork or crushed by tunnels that collapsed a couple of feet below the ground it is certainly a sad amount that was lost in the course of time. However, their efforts didn't go to waste, since it was due to the new irrigation technique that Garamantes achieved many different crops. Instead of relying upon the pork, beef and lamb raised by local farmers Their diets were coloured with their own fruit, such as figs, olives dates, millet wheat barley, sorghum and.

The first fount was there for for at least a few years. As the source exhausted and the Garamantes sought out an alternative aquifer, erected foggaras in addition, and linked them to the spring. The process continued until the Garamantes mistakenly drained what appeared to be the end of the aquifers that existed in the region, ending their existence.

It is believed that the Saharan city Carthage was another stronghold that was a co-ethnic partner with Garama. The local lore states that Carthage was first established by the Phoenician princess Dido of Tyre who was also known in the title "Queen Elissa." Dido was married to a wealthy Phoenician known as Sychaeus who won not just the respect but also the envy of his fellows Many of them were envious of his secret stash of gold. It was available in all forms imaginable including coins, bars of bullion and even massive chunks that were pure gold. Even Dido's father King Belus and her younger sister, Pygmalion even with their wealth, yearned for the treasure of Sychaeaus.

In the event that Belus was killed in 813 BCE newly-crowned Pygmalion summoned his brother-in law and demanded that he surrender his gold. Incredulous at the greed of the king's brazenness Sychaeus resisted and it could have cause him to lose his life. The desire for gold was more intense than ever before, Pygmalion removed

Sychaeus's dead body and washed away the blood from his hands and gave his sister with an untrue reason for her husband's death.

Pygmalion could have found the treasure were not due to The ghost of Sychaeus who, while seeing Dido late in the evening and revealing to her the details of his death. The ghost then gave Dido the steps to his treasure, and concerned for Dido's life and her safety, ordered her to flee Tyre immediately, taking the gold with them. Dido was overcome with emotion including anger, sadness, and awe and, as she knew she had nothing left to waste, she followed exactly what was instructed.

Destiny led Dido toward what would eventually be Carthage which is now the city on the coast located in Tunis (capital of Tunisia) situated right beside turquoise waters that flow from the Mediterranean Sea. Since the land was in the hands of a tribe of indigenous people, Dido had to negotiate her way into the territory. Utilizing all her savvy to good use in her offer, she gave the tribespeople a small portion of the gold Sychaeus had accumulated in exchange for all the property that she was able to "contain" by using the hide of one bull. The tribespeople quickly took her offer and were likely laughing all on the way home over the uninspiring offer. The smirks were quickly snuffed out the following morning when they saw bull hides carefully laid out in an elongated dashed fence which was affixed to the hill of the city and extended towards the ocean. The people of the tribe were made to feel humiliated however, since they were true to the word they bowed to Dido. She was Queen of Carthage for a long time, and created refuge for refugees, immigrants exiles, refugees, and others who were not fit for.

Christophe Cochet's image of a statue of Dido

According to historians, the legend that surrounds Carthaginian queen Dido is exactly the same. They believe they believe that Ancient Carthage was an ancient Phoenician colony that was able to reach its peak after 332 BCE, which was after the demise of Tyre by Alexander the Great. The remaining Phoenicians took their sacks with them and set off for the Tunisian coast. Since they were largely to the aristocracy they had the ability to construct powerful powerhouses quickly. The city was transformed into the beating center for the Phoenician trade within a short period of time.

Gaining the trust of close Phoenician cities such as Utica, Hippo Diarrhytus, Leptis Parva, and then, additional North African cities, including Garama and Garama - enabled the Carthaginians to establish a solid and rapid-growing economy. They also brightened the bustling Saharan marketplaces with locally grown the olive oil and fish paste as well as various spices, seasonings and other ingredients. But it was their enduring relations with merchants of in the Iberian Peninsula - in particular, Canary Island and the British Isles - that they loved most, because they were their principal source of silver and Tin. Tin, for the Carthaginians was a crucial commodity as it was required to make bronze.

The Carthaginians appeared to have been born with an exceptional perception of the direction. Highly skilled Carthaginian sailors like Hanno the Navigator faced the unknown waters and secured an exchange route that led to Ivory coast and the Gold Coast of Europe, while Himilco the Navigator made his way to the north across the Atlantic and, with good fortune as well as skill made his route to England. To ward away rivals, like the Greeks and Greeks, the Carthaginian seafarers crafted and circulated tales of sea creatures and sirens that turned their boats upside down and ate their crew members on the Mediterranean Sea. There were even absurd tales of the enchanted "killer seaweeds" floating about in the sea at the time. There was also a navy built to protect their riches from the sea however, but also to be the primary security line.

In just a few years, Carthage was one of the principal Phoenician places of authority throughout the Mediterranean. Socialites and nobility lived in castles, while people in the lower classes resided in tiny cozy cottages. It was financed through tolls and taxes from private citizens as well as business.

The lavishness was also evident in the two Artificial Carthaginian harbors, which were each defended by walls. Even though the city itself was protected by bulwarks the legitimately nervous Carthaginians compensated by building the first harbor, marked by a set of columns in the Greek style, adorned with attractive sculptures. It was the home of two warships of the Navy. The second one housed the vision-blurring fleets of merchant vessels. Additionally, it was located close to the central market of the city and was what was known to the Greeks refer to as an "agora." It was the market also served as a square for the city, where people gathered to vote and the enactment of new laws and other gatherings. The city's laws on trading were drafted and enforced by a semi-republican administration. The two democratically elected sufetes, who Greeks have compared to King-like figures and kings, were the Senate as well as the citizens' council as well as a variety of five-person boards, referred to by the name of "pentarchies."

It didn't take long for the major powers of Africa as well as Europe to become angry with the Carthaginians their influence over the Mediterranean. In a desperate search for power and the maritime monopoly that Carthaginians established within the region The Romans as well as the Greeks started fighting Carthage. The conflict between the two forces grew over the several years, with the conflict among Carthaginians and the Romans as well as the Carthaginians growing to an extent that they led to the Punic Wars.

The Punic Wars, which consisted of three major wars that lasted from 264 to 146

BCE It is believed by some to be the beginning to be the first of all world wars. Even though Carthage was defeated in its position in the First Punic War and hegemony in the Mediterranean The Carthaginians appeared to have the upper hand for a time throughout the Second Punic War. The war is most well-known because of the famous Carthaginian General Hannibal Barca who was referred to by some in the media as "one one of the more famous people that have ever lived in humankind." Hannibal, famed for his utilization of 40 elephants as their front lines of troops, crossed over the Alps and crossed the Italian borders, and defeated one Roman army after another. The end result was that Hannibal who was outnumbered 10-1, did not succeed in capturing Rome and was forced to withdraw to Carthage. The Romans pursued him and ultimately ended up winning victory in the Second Punic War.

Following the grave threat Hannibal faced throughout Hannibal's dangerous threat during the Second Punic War, the Romans did not hesitate to fight the Carthaginians during the Third Punic War, which culminated with Roman troops slamming Carthage into sand. According to legend that the Romans literally sprayed salt on the land on the ground on which Carthage stood in order to destroy it completely. While Carthage was a major effect on Mediterranean over the past five centuries, there is no evidence of the past of Carthage remain. The city itself was destroyed to ashes by the Romans who tried to eliminate any physical evidence that substantiated its existence. Even while its ruins have been discovered however, they've not offered anything like the amount of archaeological artifacts or evidence of ancient sites such as Rome, Athens, Syracuse and even Troy.

After the dust had dried after the dust had settled, the Romans took over their newly acquired territory. They dug away the rubble then rebuilt Carthage according to the terms they had set. In the 1st century, Carthage was renamed "Carthago" by its conquerors been able to become the second-largest town in western part of the Roman empire. Two centuries later after that, the Romans built two amphitheaters inside The Tunisian city El Djem. The first amphitheater built from limestone that was known as "tufa," is the most obscure of the two that was built during the the 3rd century CE. The blood, sweat and tears that went into the building of this amphitheater that was not named was not a success as it fell into disuse a few decades later.

in 238CE, known as the "Year of the Six Roman Emperors" Gordian I, one of the six emperors that were who were crowned over a period that lasted 12 months was able to order the construction of a second amphitheater which was 4.5 miles to the north of its predecessor. The plans of the freestanding oval marvel that was entirely constructed by bricks made of stone, which decided to not build foundations, were based on those in the Roman Colosseum. The Tunisian amphitheater wouldn't be any smaller than the Colosseum it is the biggest amphitheater in Rome and boasts a larger axis of 485.56 feet in all. Additionally the stone bleachers which would be placed along the walls would rise to 118 feet, with designers aiming to have the maximal capacity of 35,000 guests.

It is likely this to mean that it was the Roman Empire was now enjoying the splendor of its newly rediscovered wealth. However, the catastrophic power struggle caused the Roman Treasury at risk of a complete collapse. The struggling government was unable to maintain its financial stability, let to pay for such a lavish project. As a result when the management of the project was in a position to not pay the rising costs, the amphitheater never completed. However it was the most iconic amphitheater that became one of the city's most famous landmarks. The amphitheater, which was declared an World Heritage Site in 1979 it is still one of the most visited tourist attractions, and is so well maintained that it has been utilized as the backdrop during the 2000 Hollywood movie Gladiator.

A photo of the amphitheater's ruins.

At the time that these amphitheaters were being constructed, Saharan merchants introduced into their trading packs of animals which would transform Arabian commerce. Locals refer to them as the "ships of the desert,"" domesticated camels are often associated with desert environments were not indigenous in the Sahara. Instead, they were brought to Africa from the southeast region of Asia in the period between 3000 and 2500 BCE. Camels tend to be "individualistic," but docile creatures. They seldom traveled in groups, making them all the more easy to take. It was only with the breeding of these camels that they could learn to live together.

Camels were originally used to aid in moving people and luggage in addition to self-producing vessels for milk. Although they were handy but their numbers paled against the numbers of horses and donkeys. It was only after the consolidation of the political and military status of the Saharan Arabians that they started importing camels to the desert in large numbers, and the camels were used as desert war horses. The trainers strapped their horses using massive loads, mounted them and led them through obstacles which changed each week. They also reformulated their diets in order to improve their performance on the battlefield.

The role of Saharan camels was changed during the fifth century BCE. They were now "draft animal," they entered the agricultural field, and were changed into four-legged machines that pulled seed carts across fields and powered mills, pumped water from wells and also performed other strenuous tasks. It would take 700 years for the practice of using camels as trade animals to become popular.

Camels can not only carry up to six61 pounds when traveling long distances, but also more than 1,000 pounds on shorter distances, which makes them the ideal method of transportation for a variety of items, but they were built specifically to be used in deserts, making more efficient than horses. Camels are equipped with webbed, light feet that stopped their feet from sinking into sand, making it possible for them to move much faster. Also, the feet of camels are covered with thick soles. each heel is a huge "ball full of fat" which kept their feet cool in the hot, prickly temperature of the sand.

Camels are still admired because of their capacity to forgo water and food all of 17 days. Each time they exhale, camels retain water vapor within their nostrils. It is then absorbed by the body to self-preserve and to conserve water. Their humps, which house them with their "concentrated body fat" contain the insulation heat generated by fats in one spot which allows them to be more able to withstand temperature. Additionally, the camel's 3 eyes, bushy eyebrows and thick, doll-like eyelashes protect them from shining sun's rays.

It is believed that the Saharan merchants, who allowed the majority of their camel herd to rest from battle and then stuffed them full of dates, wheat as well as hay and grass prior to distributing their camels in caravans. A caravan was typically comprised of around 1,000 camels but there are some that are said for being as big as 12,000. With hulking cargoes filled with gold, salt silk, pepper and ivory, and occasionally pulling carts loaded with slaves' shackled cargoes Camels sped across Trans-Saharan trading routes with sand and gravel squeezing beneath the webbed footwear of their.

Holger Reineccius's photo from a Salt Caravan taken in 1985.

Chapter 3: Saharan Slaves

"Miles and miles of sand."

There is no horizon visible,

The caravan continues to move,

seeking some peace." in search of an oasis "Across across the Sahara," Gita Ashok

In around the time of the 6th century CE the growing popularity of trading caravans in the Sahara brought a brand new tribe to the forefront known as the Tuareg. Like other ancient civilizations their origins for the Tuareg are largely unknown. Some consider them to be the direct descendents of the Garamantes Some believe they are a shrewd spiritual, imaginative, and creative people that are of Egyptian descent. However, the majority is that the Tuareg can be traced back in the direction of Berbers.

The Berbers themselves were part of an ancient ethnic group which Herodotus described by the name of the "Caspians," an archaic tribespeople who settled in the southwestern and southern coastlines along the Caspian Sea. Within the Caspians was a group of people who, unhappy by the ever-growing crowded conditions of their homeland, left the homeland. They left from the region that is today Azerbaijan along with northwestern Iran and then headed west and eventually ended up on the border of Sahara. As time passed the Berbers were born in the Saharan Caspians and began appearing across Egypt, Libya, Algeria, Morocco, the Canary Islands, Niger, and Mali.

The Berbers as well as others of the Caspian tribes that emerged around the time, weren't solely of "Caspian" origin. They were actually a mix of tribes comprising people of Middle Eastern, European, and African heritage, which included the following tribes in its roster:

Romans, Carthaginians, Turks, Arabs, Vandals, Byzantines, French, Italians as well as Spaniards and all of them were ruled by Berber territories in the past. In the end, the Berbers did not have any identifiable political identity, and they were never defined by the concept of a "unified Empire," for their lands were ruled by their sovereign governments. Yet it is true that the Berbers had a distinct language known as "Tamazight. "Tamazight," now considered one of the oldest languages spoken. It was part of the vibrant Afroasiatic language family, and was able to communicate with a myriad of dialects. This is the reason that the Berbers, who are not distinct from each other, are no longer considered just a race, but as an ethnic group.

While the old Berbers were progressively different in how they governed their communities and the customs they adhered to some similar religious beliefs and passed on to their kids the stories of folklore that were handed down through their predecessors. The religious system that was prevalent among the first Berbers was based around a god pantheon that mostly focused on spirit associated with Mother Nature. Later generations added elements of local African mythologies, as being influenced by Judaism, Hellenistic religions, and Iberian beliefs.

According to the writings of ancient writers, Berbers like the majority of North African tribes, deemed rocks as god-given gifts given to them by gods. They constructed their places of worship using the most impressive boulders. They also served as platforms for human and animal sacrifices. The next passage, which was written by Herodotus gives a deeper understanding of some of the sacrifices performed in the name of Berbers: "They begin with the victim's ear and cut it off and then throw over their home. Then by killing animals by twisting its neck. They then sacrifice it to the gods of the Sun and the Moon however, they do not sacrifice it to any god other than the Sun and Moon. ..."

The most well-known of the Berber sacred stones can be found in the Mzora (also called Msoura (or Mezorah) Stone Circle in Morocco. The Mzora is one of the largest among its type, lies approximately 7 miles away from Asilah, the city of Asilah and is about 17 miles away from the hilltop Roman-Berber-Punic city Lixus. The city is situated on the outskirts of the gold Sahara the stones circle as the name suggests, is the tumulus that is encircled by a the ring of megaliths or massive, individual pieces of stone that can stand without support. In all, the stone circle contained 175 megaliths, with the highest of them creating shadows on the others at an altitude at 16 feet.

Rhaas Dyk's image of a part of the Mzora

In the first time that it was constructed, the tumulus, the ancient burial ground at the center of the circle grew to an elevation of around 20 feet, and was 157 feet in diameter. Because of the work of excavation by Spanish archaeologist Cesar Luis de Montalban during the early 1930s, little remains of the hollowed out hill, but an X-shaped mark. Unfortunately, there's little evidence to support the shape of the monument because Montalban's findings were never published and remains incomplete, but also undiscovered until today. Therefore what is known about the body or bodies laid to rest in the tomb is still not solved.

Locals in some areas took not a problem with historians' inability of determining the nature of the tumulus's owner. Based on the tales of its cursed history and terrifying tales of researchers going insane that were raging in the aftermath of Montalban's disappearance public eye during the dictatorship of the fascists, they preferred to keep it that way. Others, just like their ancestors, were fascinated by the mystery and believed it was an unimaginably beautiful monument that was sent down from heaven. Every grain of dirt, sand and gravel was thought to have supernatural abilities, which led some even to assault the spot by scooping up as as much of the holy dirt as they were able to before disappearing. However, the hill hasn't ever remained untouched and legend says that every time the cavity-ridden mountain appeared on the verge of collapsing, massive rains fell from the sky until the hill was a complete again.

Farmers who needed plentiful rains were believed to be one of the main causes.

Another theory, which some believe is attributed to the famous historian Plutarch is that the remains of the mythical Libyan gigantic, Antaeus, that once was buried beneath the surface of the tumulus. Antaeus was believed to be the unstoppable offspring of the Greek god of the sea, Poseidon and his wife, the Earth goddess Gaia (or Ge). The giant of cocksure was a swaggering figure at the entrances to Berber areas in order to challenge anyone who wanted to enter to a match of wrestling. With unbeatable power, Antaeus defeated all those who were foolish enough to accept his invitation until he was able to meet Hercules.

Hercules was the god of Zeus has killed numerous beasts and foes in his lifetime, and as Antaeus his record was flawless. The rivals at first were akin in physical strength, fought to defeat each one.

Every time, Hercules flung his rival to the ground and Antaeus bounced back ready to strike with a more devastating strike. As it dawned on Hercules that Antaeus was the son of the goddess of earth and was charging his power every time he fell to the floor, he threw the fumbling giant off the ground and slung the head of Antaeus until the last amount of power was drained from his body. He was triumphant, Hercules dropped Antaeus the limp body to the side and then brushed his hands and headed home. The Djouhalas - an enigmatic bunch made up of "pagan giants" who later discovered Antaeus carried him to the spot that the tumulus stands today and laid him to rest there.

It was not until the 9th century CE that Berbers began to break away from their long-standing traditions and accept the teachings of Allah. The word "Tuareg" was invented by the Crusaders and means "abandoned in the hands of God." Some also attached to them the nickname "the Blue Men of the Sahara," as the distinctive indigo scarves they wore frequently threw off some blue. When it came to the time of their death there was no need to worry about the opinions of those who weren't familiar to their customs were thinking about them, since the hulking tribespeople were familiar with their own identities. They referred to themselves as Imohag, also known as "free people."

The exact date of the Tuareg's arrival has yet be established, but the majority of experts believe the Tuareg and the Berber caravan tradesmen, initially showed up in Fazzan and were enticed with the unclaimed grass land the region provided. It was only after what historians refer to as"the "Tuareg diaspora" that they began appearing in other parts of the Sahara like Niger, Algeria, Mali in addition to Burkina Faso. They quickly created a Berber-inspired language that they named"the "Tamachek," and even created their own script that was influenced by Libya called"the "Tifinagh."

At first of the century, people of the "semi-nomadic" Tuareg lived separately within various clans in the Sahara. Each village was made from spartan, collapsed houses that were joined by weaved mats, wooden frames and spare fabric. Each was administered by the tribal system of governance till the wise and strikingly gorgeous Tin Hinan, her bronze skin as radiant as the sun's midday rays, set out on an "milk-white camel" along with her servant Takamet who was within the mountains of Algeria. Tin Hinan was the "matriarchal mother from Tuareg," Tuareg," who consolidated all of them into one kingdom during the fourth century of CE. She then was crowned their first queen.

In the mid-600s, "pastoral Berbers" began to leave the city limits in a masse. Two clans with a goal to improve their lives was those of Lemta as well as the

Zarawa which were close relatives of many of today's Tuareg.

The structure of early Tuaregian societies was based on feudalism, with every participant, based on his or her social status, either blessed or fated to lead the lives they were assigned to the members at birth. At the top in this pyramid of power was Tuaregian nobility. In addition to the fact that these wealthy tribesmen have the highest number of camels, but they were also the ones who ruled over everything in the Tuaregian caravan business. But their homes were not as luxurious as the huts that were the homes of people of Tuaregian upper class as they spent the majority of their lives in the open. All through the year, they traveled through their Saharan areas, stopping at each oasis town to ask for the harvest share and other tariffs and taxes that they owed.

The most interestingly unorthodox feature of the surprisingly modern

Tuareg was the exact feature that caused others North African tribes to shun their members. Simply put, the older tribes, especially those that were part of the Islam faith, mocked the role for women within Tuaregian society. The matriarchal norms society allowed Muslim women to stroll all around with their beautiful faces unveiled, while men were required to cover their face with blue scarves, which left only a an apex for their eyes. The most shocking thing to anyone outside was the women's ability to be sexually active, since they didn't have to be virgins prior to marriage and were not punished because of the fact that they had more than one partner. Women were (and are) legally entitled to all the husband's assets should their marriage break up with divorce.

While Tuaregian women certainly enjoyed privileges and rights that the majority of women in the past could only dream about however, their conception of equality different from what most believe to be a fair definition of equality today. Although women were able to be granted inheritance and buy their own camels, if they could be able to afford it, as an example but the task of herding camels was normally a task for males. Women were certainly valuable social members however, they were viewed as fragile and fragile and were entrusted with the care of donkeys, goats, sheep and others "timid" livestock.

The second reason was that women were completely excluded from caravans. The only option available to them was to hire an male relative or hired employee to work in the trade for their benefit. Male representatives, who financed the caravan together with the camels of their employers were required to return with either the goods they requested or any profits made from the sale that were later divided equally between them.

Like many of the North African tribes at the time, the strength of the Tuaregian economy was entirely dependent on their success during trading in Trans-Saharan trade. One of the five important Saharan trade routes that linked cities along the Mediterranean coast to the cities on the southern end of the vast desert was a west-facing circuit known as"Taghaza Trail. "Taghaza trail." The path started in the north of the city of Aoudaghost (close to the present-day city of Fez within Morocco) and slithered across Sidjilmasa, Taghaza, Walata, Ghana, Bamako, and Niani before settling in the area that is now known as Sierra Leone's Freetown located on Freetown on the Atlantic coast. Caravans that traveled south brought along crude glass oil lamps, ceramics beads, saffron dates, flour, and. Merchants travelling north carried containers of nuts from kola as well as slave-carrying wagons. The ones who took the west-bound route from Idjil to Walata and brought salt blocks.

Due to their nomadic heritage, in particular their mobility and unbeatable endurance in scorching heat, they took the chance to become "middlemen" for this Trans-Saharan trade. After they received the cargo from their customers then they set out for their destination on the Mediterranean coast. When they arrived they transferred the merchandise to merchants from the maritime sector who proceeded to sell the goods internationally.

It was said that the Tuareg Deliverymen became so sought-after that they quickly could refuse opportunities for business. Due to the small space available in the caravans, along with profits and efficiency in mind they chose only customers who dealt in expensive (and often small) high-end goods. Many of these traders have been granted the opportunity to travel outside their own borders, eventually settled in the villages and towns they discovered along their trade routes. Some of them worked in an artisanal local business while others were ambassadors to their nomadic sisters and brothers.

At the time salt was regarded as an element of "white golden" in the Sahara. Apart from giving flavor to their meals it was used to keep meats and other perishable items. To transport along the maximum amount of salt at once, independent Tuaregian merchants built huge caravans of 2,000 to 4,000 camels that they were referred to as "Azalay." On arriving at salt mining sites laborers, slaves, and merchants themselves pounded for salt. Each piece of salt was first submerged with mud, to "protect them against the weather" before being taken into the caravans. When they had accumulated the required amount of salt, they began to travel to home. Along the route, they stopped in various oasis towns to exchange goods with locals, trading bars of salt in exchange for drinkable water, dairy products, bracelets, meat as well as textiles dried herbs, and other things of value.

In the end, as salt became increasingly accessible and affordable, its price took a decline. Instead, there was a renewed and increasing demand from African slaves. Then, Tuaregian merchants geared their business accordingly. Many chroniclers believe that the initial Tuareg's brutal treatment of slaves was largely to be related to the negative connotations associated with their name. Some attribute the negative perception of Tuaregs due to "swarms" of Tuaregian highwaymen and pirates that terrified innocent people and foreigners alike. However, there is no proof that such bad apples were commonplace, and many reports of traders who had contact with the Tuareg in person remember they were a peace-loving and cooperative group.

Though they were not considered to be violent but the Tuareg were combatants who were equipped to defend themselves should it they needed to. Some were believed to strangle their enemies to death using their scarfs, while some chose to take on their foes by using broad swords with round tips known as"takouba. "takouba." The sparkling blades, that averaged in between the 29-33 inches long were masterpieces in their own right, sporting intricate etchings on the base, as well as their leather or brass cross-guards as well as scabbards that were stamped with the eight-pointed al-Quds Star and various other Islamic symbols.

Another method in which the Tuareg protected themselves was through the introduction of the ksour. A ksar is an attractive fortified city that is built on top of oasis, built entirely of adobe and cut stone. They are surrounded by tall clay walls, and adorned by "orthogonal roads" and one gate that faces towards the east. In order to increase the area of land that is fertile the buildings within the ksour, some of them up to four stories high like dome-shaped private houses are built on top of each other and accessible via staircases that zigzag on the façade. There is also mosques, granaries, bathrooms, a communal oven and a public square and shops within these Ksour.

Ruins of a ksar's ruins in the present day

Algeria

Jerzy Strzelecki's photo from Ksar Ait Benhaddou from Morocco

A different Saharan city that accelerated it into the forefront of the ever-changing Trans-Saharan trading was Timbuktu. The city is located within Mali within the southerly part of the Sahara, just 12 miles to the north of Niger River, Timbuktu is described as Robert Launay, an anthropology professor at Northwestern University, puts i as the "port of entry into the desert into North Africa." Scott Neuman is a journalist for National Public Radio, explained, "For centuries, it was a major trading hub that connected Europe as well as its neighbors in the Middle East, and later an outpost linking with the West African coast with the Sahara's interior, which is largely unexplored." Alongside its strategically placed geographical location Many took into account the abundant natural resources and stunning landscape of Timbuktu described it as "the "Jewel of the Sahara," and some consider it to be among "the "great old wonders in the universe."

A 19th century representation of Timbuktu

It was around 1100 when Timbuktu was an official settlement. When the trade routes of the past were disrupted by foreign competition The Timbuktuese were quick to seize the opportunity to become a "transport hub" for caravans that were passing through. Two centuries later the city was taken over by the Mali Empire. It was no longer a place that was primarily known for its trade prowess and its trade prowess, it became a thriving "cultural mecca" with talented artisans, scholars, crafters and other outstanding minds. The library, which held numerous Islamic, African, Greek, Roman, and other unique manuscripts, as well the stunning three mosques, placed Timbuktu in the spotlight to be one of the top sought-after educational centers in the world. When the number of universities and schools in the city grew the number of students and young scholars from all over the world came to

Timbuktu and many were awarded the scholarships offered by these prestigious institutions.

In the 1500s, when Timbuktu was part belonging to the Songhay Empire and had a population of 25,000, more than 25000 people, which is about 25 percent of the population were enrolled in school. Scribes from the religions made a livelihood by writing and copying significant Islamic manuscripts, while secular scholars gathered, studied and re-examined the writings of the great minds working in fields like maths as well as law, astronomy and geography and world geography. At one time, this literature was so valuable , many believed that it was a money. Leo Africanus, a 15th century writer from Berber-Andulasi heritage wrote "In Timbuktu, there are many judges, scholars and priests. All paid well by the king, who lavishly honors scholars. A large number of manuscripts directly from Barbary can be sold. These sales can be more lucrative than other kind of product."

The Medieval Timbuktu was also brimming with other valuable products including gold and salt. Europeans were especially enticed by the dazzling troves of Dhahab, which were hidden within the city. The moment Mansa Musa, King of Mali who was in search of these gorgeous reserves in 1324, the king sucked these mines with such fervor and energy that within the time he was crowned the price of gold plummeted that was not able to recover for a number of years.

When Timbuktu as well as the cities surrounding it began to focus to the slavery trade, the idea of the business had existed for many tens of many years. The terrible business that would not die out until the 18th century in the words of UNESCO along with the Nigerian Center for Black and African Arts and Civilization (CBAAC), "unprecedented in history and was the cause of the deportation of millions Africans to different parts of the globe."

It was a 7th-century Egyptian King, Abdallah Ben Said, who encouraged the Trans-Saharan slave trade. After completing his capture of Sudan in the year 652 and referring to it as"the "land for the false prophets" Abdallah wrung the arm of Khalidurat the Sudanese ruler and finally signed on to the novel treaty known as "Bakht." According to the terms set forth in one of the clauses in the treaty the king was required to provide Abdallah by supplying him with at minimum 300 African slaves per year. The present-day phenomenon called"the Transatlantic Slave Trade, fueled by the increasing supply of African slaves from the Sahara and the Saharan Gulf, began to flourish towards the end of the 17th century, and the world is getting towards an Industrial Revolution. In addition to the growing demands for local manual workers, traders of slaves from the

Americas as well as in the Middle East, the Persian Gulf and even regions as far away as China and Japan have bought unending shipsloads of slaves from Saharan capitals and kingdoms to use for their mines and plantations. Tidiane N'Diaye is an expert Sengalese Anthropologist and the writer of Le Genocide Viole, reckons that 10 million Africans suffered from this reprehensible method of killing.

Chapter 4: The Modern Sahara

"I am a desert lover. I love the vast wasteland in the mirror-like trembling of the fata Morgana, the wild rugged peaks, and the dune chains reminiscent of the stiff waveforms of ocean. And I am awed by the simple, rough life of a primitive camp during the star-lit, icy night and the hot standstorms too. ..." A short excerpts from "The Unknown Sahara,"" translated by Z Torok.

In 1923, Oxford-educated Ahmed Hassanein saddled up his camel, set off to Sollum in Egypt which was the first person to take on the entire eastern portion of the Sahara which is known as it's the Libyan Desert. After a long and strenuous 8 months Hassanein was able to travel 2200 miles and gained another place in history by revealing the forgotten oasis that were Arkenu and Uweinat He described them in terms of "rising as medieval castles of the sand of the desert," in the southwestern crevices of Egypt. He meticulously removed and revealed paintings of animals like gazelles, ostriches, giraffes and lions, as well as cow-like creaturesthat were buried in the rocks. These discoveries that he recorded using his best efforts in dark black and white photographs that were grainy, showed evidence of human presence in these areas which dates back to at least 10,000 years.

Hassanein

As one would imagine, Hassanein was nearly blinded by the media attention when he returned to Sollum. He was immediately contacted from American investors who offered before him an amount of $20,000 (approximately $260,872 in the present) to lead a tour across in the United States for a couple of months. The American public due to the press were familiar with his name by the name of "Egypt's Lawrence of Arabia," and was then instructed to drape it over his heavy and wear his most "authentic" Bedouin robes. To their dismay, Hassanein demurred, including by refusing to make the following assertion: "My standing precludes earning money in this manner."

In August of the same year, Hassanein organized in Alexandria an extravagant gathering that was with a host of African princes, ministers government officials as well as doctors, scientists, and other learned men of high rank in the world of. In front of a podium dressed in formal attire, Hassanein presented his majesties, superiors and friends the specifics of his journey. He then presented an illustration with the exact locations of each town and city - including Arkenu and Uweinat that he been through during his travels. This was how Hassanein was the first person to draw a precise diagram that covered The Western Sahara.

Mohamed Mabrouk's map that is based on the Mabrouk's map based on the

About 30 years later "roughnecks" located in the southern part of Libya who were obsessed with exploring the earth for hidden pouches of oil discovered an untapped aquifer beneath the Libyan desert. Locals who were grateful for the finding, since it came at a more appropriate time. the soaring salinity levels, and the contamination from the rusting pipes in the ancient aquifer of the capital city, Tripoli and made the water unfit for consumption. Although the idea for this new aquifer infrastructure was designed to improve the irrigation system and provide water to the Libyan population of 6300,000 was initially conceived in the 60s and was put in the year 1983. In 2011, three of the five phases of construction for the massive undertaking, referred to in the "Great Man-Made Project" are completed. Engineers plan to build pipelines with 284 miles, and create 1,100 wells connected to five reservoirs capable of releasing 6.5 cubic meters of water per day. Despite the high cost for the project, estimated at around $25 billion dollars and the amount of "non-renewable liquid" is just 10 percent of the amount it would cost to purify salt and other nagging minerals from Tripoli's salinated waters.

The majority of Saharan states gained their independence following the Second World War. Today, the Sahara desert hosts around 4 million people who reside within Egypt, Algeria, Libya, Mauritania, and Western Sahara. There are more than 500 plant species and at least 250 animals. Alongside the many nomadic tribes who continue to wander around the Sahara present day are urban and village dwellers that rely on their oases in addition to the abundance of agricultural harvests, and the local mine industries to provide food.

A few of the mystery of Sahara are yet to be discovered. On the 8th of January the 8th of January, 2018, photos showing the dunes in the desert, covered with lacy-white snow that was that were as deep as 15 inches were shared on social media overnight. The one-inch snow layer that blanketed on the city of Aif Sefra, known to locals as the "gateway to the desert" marks the first snowfall that has occurred in the town in more than 40 years. It's yet another reminder of which age Sahara is and regardless of how far back the existence of human beings is it is clear that it is a place that "Great Desert" can still deliver unexpected events.

Chapter 5: Water

The Sahara has just two permanent rivers, and several lakes, however it is home to large underground reservoirs or Aquifers. The two rivers that are permanent to the Sahara are the Nile and Niger. The Nile is located in central Africa just north of the Sahara and flows northwards into Sudan as well as Egypt and then empties into the Mediterranean. It is the Niger rises in the western part of Africa south-west of the Sahara and flows northeastward through Mali as well as the desert. It. It then changes direction to the southeast through Nigeria and eventually drains to it's Gulf of Guinea.

The Sahara is home to around 20 lakes however, only one of them has water that is potable - the vast, but very shallow Lake Chad, a continually expanding and shrinking pool of water located in the region of Chad located in the southernmost part of Sahara. Some lakes contain a sour stew of water that is not drinkable.

The Sahara's aquifers typically lie beneath an intermittent drainage also known as "wadis," which rise in mountain ranges and drain into the floor of desert. Aquifers can discharge a portion of their water into the desert floor at places known as "oases," which are usually found in lower regions of surface depressions.

Climate

The Sahara is one of the most extreme climates. The Sahara area experiences very little to no rain with powerful and unpredictable temperatures, and a wide range of winds.

Over across the deserts, average annual rainfall can be as low as several inches or less lower in some locations. In some regions where there is no rain, it can fall over a long period of. After that, several inches of rain could be sprayed down in torrential downpours. After that the rain will cease completely. will fall for several years.

The dominant winds, which blow from the northeast towards the equator all year, is responsible for the desert's dryness. When the wind shifts toward the southwest, the air heats and disperses moisture that could otherwise be released as rain. Locally, hot winds frequently bring dust and sand out of the ground, and spin them upwards through the cooler air like dust devils, or propelling them towards the southwest in intense and violent dust storms that blind.

In the summermonths, daytime temperatures in the Sahara can often rise over 100 degrees Fahrenheit with the highest air temperatures meteorologists have ever recorded (136 degrees)that occurred in Azizia, Libya, on 13 September 1922. With blue skies, temperatures can drop 40 degrees or more in the evening. In winter, frigid temperatures could be observed in northern Sahara and cooler temperatures can be found in the south of Sahara. There are occasions when snow falls in certain higher mountains, but not often, on the desert floor.

Wildlife

The Sahara's harsh environment demands that animals adjust to extreme drought as well as fierce winds, extreme temperatures, and wide swings in temperature. In the middle of the Sahara For instance, the majority of mammals are tiny that helps reduce losses of water. They usually get their water by eating. They shelter inside burrows in the daytime, hunting and foraging mostly in the evening in cooler temperatures. They have evolved anatomical adaptations , like the fennec's ears that are large that help to disperse heat and also its floppy soles, which safeguard its feet.

The Sahara is home to more than 70 mammal species and 90 species of birds that live there and reptiles, as well as 100 species as well as a variety of arthropods (invertebrates that have joined legs, segmented bodies, as well as exterior skeletons). The animals include, to name just a few instances, Barbary sheep, oryx anubis baboon and dama gazelle, spotted hyena common jackal, and Sand fox; the birds include ostriches and secretary birds, Nubian bustards and various reptiles like cobras skinks, chameleons, numerous lizards, as well as (where there is adequate water) Crocodiles. Also, the arthropods: scarab beetles, numerous ants as well as scorpions, scarab beetles as well as the "deathstalker" scorpion. The wildlife is found primarily on the less harsh northern and southern borders and around desert water sources.

The Sahara's most well-known species is the dromedary, which has been domesticated for thousands of years and extensively used by desert nomads. With its fat-filled hump as well as other adaptations to its physiological environment the dromedary is able to wander for days without water or food; and with its huge lips that are thick and a wide mouth, it is able to eat tall, thorny plants and salt-laden vegetation and dry grasses. With its large feetpads, it is able to traverse dirt and sandy terrains; with its slit nostrils , thick lashes and eyebrows they can shield their eyes and nose from the ravages of storms of sand; and, when provided with water, it is able to consume up to 30 Gallons in just minutes, readying itself for hot, dry days.

Plants

Like other deserts Like all deserts Sahara has a comparatively small amount of plants that are wild With the greatest concentrations being found in the northern and southern edges and close to the drainages and oases. It has forced changes to the plants. For example, in the vicinity of wadis and oases, species like date palms, tamarisks and acacia grow long roots in order to get water. In areas that are more dry flowers of plants grow quickly following the rain, placing small roots and then completing their cycle of growth and producing seeds in just a few days, just before the soil begins to dry out. The seeds can remain dormant in the soil that is dry for a long time, waiting for the next rain to begin the process again.

In the most arid areas like the southern region of the Algerian Tanezrouft Basin, a fearsome collection of salt flats, sandstones, and sand dunes referred to in the "Land of the Terror" The plants have managed to only establish the weakest foothold and have left a large portion of the landscape largely barren.

In the central, most deserted part of the Sahara The plant community is estimated to include 500 species. In contrast, in the South American Amazonian forest --

possibly the most biodiverse region on earth -- the plant community is according to estimates of some experts around 40000 species.

Chapter 6: People And Cultures

As per estimates of the Sahara's complete population could be as low as two million, which includes those who reside in permanent communities close to water sources as well as people who move from place to another according to the season and those who travel the trade routes of the past as nomadic nomads. The majority of them are of Berber or Arabic roots. The Berbers have spoken several varieties that are part of Berber, or the Berber language, came onto the scene at the beginning of Sahara's history.

The Arabs were speaking Arabic which is an Semitic language, which originated in Arabia and came onto the scene a few thousand years before. Most of Sahara's inhabitants adhere to the Islamic religion, which was introduced around the seventh century AD.

The Sahara's story is told in terms of hunting and gathering in the primitive times trading, nomadic and development of agriculture early communities, conquests advanced civilizations, massive architecture, dynasties exploration, colonization, and war. It bears the stamp not just of the Berbers and the early Arabs however, it also includes Egyptians, Nubians, Phoenicians, Greeks and Romans. In the past it was a victim on the back of Ottoman, Spanish, Italian, French and English colonialism. In the 19th century, it was able to hear the whispers about Roman Catholicism. In World War II, it was a victim of a brutal and bloody war with both the Germans along with the Allies. Then, in the middle of the 20th century, its nations threw away their colonial yokes, and embraced the freedom they had always wanted.

Wonders

The Sahara is a natural wonder with its many beauty and culture is an unforgettable trip. For a few instances to consider, here are some:

Explore oases, dune fields and along the Nile as well as the Niger rivers, and some of the poorest parts, (for instance, the Tanezrouft Basin - the Land of Terror).

Discover exotic wildlife such as those of Barbary Oryx, sheep the hyena, the jackal, and the Sand fox among others. diverse reptiles and birds.

Participate in hikes and camel treks and camel treks, bringing back the nomadic trade caravans.

Explore breathtaking sites that tell the story of humanity in the Sahara For instance the ruined remains of ancient civilizations and the modern-day edifices that represent civilizations.

Explore the varied cuisine of old, but still lively market places, bazaars and markets.

If you've never been to the Sahara and aren't familiar with local conventions and norms, you must seek out a travel agent who can provide the necessary information that you require for an enjoyable excursion.

Interesting facts

The ancient Egyptians were believed to hold the scarab with reverence since the baby of the insect seemed to appear on its own like a miracle. (In fact, the baby was born from the dung of an animal in the same place in which the female beetle put her eggs.)

Berber and Arab nomads traveled in their camel caravans across the Sahara trading in commodities like salt from the desert, gold and slaves.

The tiny stretch of desert on the Atlantic coast is home to a variety of plants, lichens, and succulents. The organisms get their water from the fogs generated from the cold Canary Current, which parallels the coast and is located just offshore.

These crescent-shaped dunes pushed by wind, can be able to travel for miles over the course of the year.

The Sahara which covers 3.5 million sq miles of land, is considered to be the most "hot" desert in the world. However the Antarctica which covers 5.4 million sq miles is by far the biggest desert. (While the Sahara is averaging only a few inches of rainfall each year, Antarctica gets only slightly more.)

Saharan travel

A trip to the Sahara is as diverse like the Sahara itself however there are certain quintesential Saharan experiences. They usually begin in cities that are the gateway to where excursions into the desert are planned. These towns are often oases. They are the hubs in Saharan culture, where the structures emerge naturally from the soil and the rate of development has not changed much over the centuries.

If you want to explore the area beyond the village boundaries or town limits There are two main ways to travel. A slow , steady walk along the sands riding an animal reenacts the legendary comel rides of Saharan legends. On the camel safari, people slow down to a speed that is suitable for Sahara's harsh conditions, allowing you to be aware of the finer details while traversing this stunning landscape at one with the surrounding. On a four-wheel-drive adventure travellers can move further, stirring the sand while ticking off the iconic Saharan landscapes.

The feeling of sleeping in a room with four walls is a unique experience that can be ended in towns. In the sands of distant Saharan mountains the evenings are spent by an open fire and a comfortable mattress of sand is a nighttime mattress that is most popular.

The majority of Saharan excursions include tents however, many travelers prefer sleeping outdoors under the most beautiful night sky on the planet.

When should you go

Traveling to the Sahara is most enjoyable between the months of October and April, or in the early part of May, during the time that daytime temperatures are manageable. Then, in the middle of Saharan winter (especially January and December) the nighttime temperatures could dip below freezing. Sand storms can be a possibility during the months of January to May, and no one should venture into the intense flames of heat that cover all of the Sahara from June through the beginning of September. The rain isn't a big issue.

Where to where to

Following Saharan countries are open to travelers and can be a good option for people who are new to Saharan travel.

Morocco

Southeastern Morocco, in the shadow of snow-capped High Atlas Mountains, is the most accessible part in the Sahara. The region is in Draa Valley - a picturesque paradise of sprawling palm groves, earthy red Kasbahs, and Berber Hamlets that the trans-Saharan camel caravans started and ended their 52-day trip through the Sahara to Timbuktu. Today shorter excursions with camels head out of M'Hamid to the Erg Chigaga, a stunning 40km-long stretch of stunning dunes of sand. Further to the east, starting from the tiny town of Merzouga the camel safaris and 4WD tours lead towards Erg Chebbi, a glorious collection of seemingly endless dunes. They both Merzouga and M'Hamid can be reached on a one-day bus ride from Marrakesh.

Tunisia

The south of Tunisia cuts a long cut into the northern Sahara creating what may be the Sahara's most famous area. It was in this region that filmmakers discovered enough cinematic beauty to be the background to scenes from the Star Wars series and the English Patient. The two major cities with gateways are Tozeur tozeur, which is a seven-hour train journey or a one-hour flight from Tunis and Douz the town of Douz, which is a nine-hour train journey in Tunisia's capital. Douz is situated close to the huge salt lakes in Chott el-Jerid and some of the most beautiful locations from Star Wars filming; the second is the final destination before the enthralling sandy peaks that make up the Grand Erg Oriental, one of the largest oceans of sand which flows into Algeria. To experience authentic Saharan immersion the isolated outpost in Ksar Ghilane, 147km south of Douz is home to an abandoned fortress as well as the Sahara's most beautiful Tunisian sceneries.

Egypt

The western part of Egypt can be described as a huge and fascinating part of Sahara. There are oases linked by barely discernible tracks in the sand, starts in the north, at Siwa which is home to a crumbling medieval fortress of mud and a temple dating to the times that of Alexander the Great. In the south the oasis comprised of Al-Kharga, Dakhla, Farafra and Bahariya are the locations for explorations to the desert, and beyond the final outpost of human settlements are locations that evoke Saharan desire, such as those of the White Desert, the Black Desert and the Crystal Mountain. Even more remote than those and Gilf Kebir Plateau is the most remote. Gilf Kebir Plateau is the most rumored home of the elusive Zerzura Oasis, and the most disputed site that is the Cave of Swimmers, made famous in the silver screen of The English Patient. The oases all have direct bus service to and from Cairo that take at most a full day to reach. Al-Kharga also has more rapid air connection to Cairo, Egypt's capital. In the oases, 4WD excursions are more frequent than camel rides.

Sahara's most stunning landscapes

The most stunning landscapes of the Sahara are located - perhaps not surprising in the desert's heart. Within these regions, Algeria has sand seas with the biggest dunes of sand on the planet and the rocky landscape of southeast of the country - and that of the Ahaggar and Tassili mountains in particular - creates some amazing Sahara landscapes. The rock art that was created by the ancients of this latter area makes it an Unesco World Heritage Site. Nearby to Libya, the Jebel Acacus is also stunning with its vast sand beaches as well as isolated black-sand volcanoes, and stunning lakes are among the Sahara's most gorgeous areas. In addition there are there are the Tibesti Mountains in northern Chad and The Niger Air Massif and Tenere sands along with

Mauritania's historic cities and lush oases and Mali's Timbuktu complete an impressive collection of Saharan destinations.

Chapter 7: What To Do Within The Sahara Desert

There's no doubt that the Sahara Desert is, without doubt, one of the most beautiful places to do various fascinating things. It might appear as if there's only the sky and sand but there's a huge array of things you can do within the Sahara Desert. In any case, what is there to want from the desert? If you're planning an excursion to northern Africa but the Sahara isn't in your plans we urge for you to reconsider. We'd suggest cutting back your plans to Marrakech, Fes or Tunis so that you can stay longer than two nights in the Sahara desert.

The desert doesn't possess any distinctive attractions, aside from the obvious dunes of sand. The desert itself is the main attraction. It's simply magical. This is a truly magical place. Sahara Desert is one of the most authentic places you can explore in north Africa.

There are people who try to earn money from visitors, but their way lives aren't found anywhere other than in Sahara. The towns in the desert are dusty, compact , and basic in their facilities, but they attract the eye like no place elsewhere.

Walking through a desert town or the dunes are the ideal way to see it and possibly be captivated by it like we did.

You can sleep in the Sahara The desert of dreams

The desert was a dream and under the stars of a million! I can't even describe the experience, and I'm not sure I can give it the full justice. It's a must. The view of the sunset and sunrise from a tent camp will make you awestruck of the beauty and natural beauty of the most beautiful Sahara Desert in the globe. Planning a trip to the Sahara Desert in Morocco for instance is simple, especially if do not have your own transportation. When you've got it, then what do you have to lose?

Enjoy an excursion with a camel

It's not unusual to ride a camel to the campsite you'll be staying at when you're on an organized tour. Camels aren't exactly the most pleasant animal to ride and the guide will always be walking along on the train. However, it's something you should do at least once in your lifetime. When you reach the desert there's no shortage of people who are willing the chance to take you on a walk. This is why it's much cheaper to schedule your trek at the time you arrive. Be sure to negotiate. Hard!

Take a quad bike ride through the dunes

When you're in your vehicle you'll discover it's almost impossible to drive large trucks or heavy-loaded 4wds up to the top of the most sand dunes. Renting a light and powerful quad is a great option to have fun in the Sahara dunes. 1

- and 2-hour trips are easy to arrange, however they can be costly. Of course, a lot of negotiation with tour guides is to be expected. Be careful though. We know of a person who died in an accident in 2014 in Erg Chebbi. We saw a man wearing a back brace as a result of an accident three months earlier. Take it easy. Dunes can appear soft and soft, but you'll end up in a awkward position with a speed that is high and you're in trouble.

Sand surfing

There are many chances to rent sand/snowboards and sand skis on the dunes, which are heavily populated. It's fun to try it at least once, but there's no ski lifts which means you'll have to return up the dune every time. To increase the performance of your ski, rub candle wax onto the base. If that fails, you can make use of a shovel or rope.

Sahara spa day

Who'd have thought that one of the most popular things to accomplish within the Sahara Desert would be to enjoy a day at the spa? It's not your average spa however! Merzouga people are from Morocco for instance, swear by burying themselves all the way to their necks with Sahara sand, as it's believed to benefit muscles and sore muscles and limbs. When it is hot in summer, particularly at the period of mid-afternoon you'll find locals digging themselves up in sand. They say it cleanses the body of ailments and , in particular, arthritis. Be aware that in the summer months, temperatures can be well over 50degc , making it an ideal place to be any time. Be aware that you should not stay in the ground for more than 15 minutes because it could cause you to burn to death. In fact, we've been told of victims who that have experienced this!

Join a convoy

We typically travel on our own, but getting together with fellow travelers and travelling together is a very satisfying experience.

Meet the locals

You'll have no options to make. You may think that you're on your own, but Berber people, fossil sellers , and camel tour operators will come to you! It's an excellent opportunity to meet and gain an insight into the daily lives of the Berber people living in the desert.

The campfire is a perfect place to relax.

You might be surprised to learn that there's lots of wood dead within the Sahara. Dead bushes make excellent fuel for a fire and will help to start a fire in a short time. Find enough wood, make the fire pit, then place your grill on the glowing embers and begin cooking! It's like nothing else and you'll never cook your meal that you've prepared in the microwave ever again!

It can get extremely frigid in deserts at night particularly in winter. the campfire is an excellent option to stay warm.

The top tip: We've learned that using a shovel filled with coals that have been thrown from the flame beneath your seat is an amazing option you can keep warm!

Spot wildlife

After you've had your campfire and dinner meal, you can leave the leftovers in your camp , and then look out to see if wildlife around you appear to eat a late-night snack. Desert mice are more likely to be first to show up and , if you're lucky enough the shy desert fox might come by to take a look. You can do this over a number of night and it will get more courageous and provide you with the perfect photograph possibility. If you're lucky enough you might even see some of the very last Dorcas gazelles. They're believed to be extinct, however we know for certain that they're not.

Stargazing

Of all the things you can take part in within the Sahara Desert, this is my personal preferred. It has a million stars over you, and with no light pollution to mention the night sky over the Sahara Desert will leave you amazed. After being through the desert for months discovering my own sand dunes in the dark, and looking up in the sky at night is exactly what I'm looking for! Space heaven for me!

Take a look at the sunset

Take a stroll along the dunes after sunset or even early in the morning to witness the stunning and breathtaking change in shades that the dunes display. With the peace and peace, the stunning array of colors will astonish you and leave an unforgettable impression on your mind. Out of all the things you can experience within Sahara Desert, the Sahara Desert, this experience will always, in a tiny way, make you want to come back one day to this amazing landscape.

Chapter 8: Antarctica

Antarctica, a desert? Aren't all deserts meant to have hot temperatures? Actually, the truth is that this is a popular misconception as there exist deserts with cold temperatures. Antarctica as the coldest spot on earth is not just cold. The coldness of Antarctica makes it difficult for the majority of species to live there. Along with being the coldest region on the planet, Antarctica is also the most windy and dry. It is considered to be dry and is classified as a desert since it only receives 8 inches of rain each year. Eight inches of precipitation is mostly along the coast, and inland areas are less humid.

The Antarctic Ice Sheet covers about 98 percent from the continental area of Antarctica. A few of the intriguing aspects of Antarctica is that even though it is a desert because of its little participation in the glacier (which is 90%

of all glaciers in the world) contains 70% of all freshwater that exists in the world!

Sahara Desert

It is said that the Sahara desert is perhaps one of the most popular anywhere in the world. when asked to identify a desert off your head, you'd likely imagine the Sahara. It is the Sahara Desert is a subtropical (hot) desert. Its name Sahara is almost identical to the word "desert". In reality Sahara is a desert. Sahara is home to all the features we imagine when we think of the desert, such as the scorching heat during the day, as well as tons of sand. The Sahara is situated on the continent of Africa particularly within the nations of Tunisia, Sudan, Niger, Morocco, Mauitania, Mali, Libya, Chad, Egypt, and Algeria. As you've probably realized, as with all deserts it is the Sahara is a difficult environment that is difficult for a variety of species to endure. The areas that lie in the middle that lie within the desert in contrast to areas closer to the edges, are most difficult to survive in.

As you consider the Sahara and the Sahara desert, you may imagine camels too. You might be shocked to find out that camels had not always lived in the Sahara in the first place, but were introduced by humans to work animals around 200 A.D. Camels are very adept at surviving the harsh desert conditions for extended periods of time, since they can easily and swiftly move through the sand. They can last more than 17 days without food or water.

Other species that thrive in the Sahara's harsh conditions are the numerous rodents scorpions as well as snakes (which we believe consume rodents).

Arabian Desert

The Arabian Desert, which is an subtropical (hot) desert that covers the entire portion that is part of the Arabian Peninsula. Saudi Arabia contains a large portion in the Arabian Desert. Other countries with a part of the desert are Yemen, United Arab Emirates, Qatar, Oman, Kuwait, Iraq, and Jordan. Rub'al-Khali is among the largest uninterrupted areas of sand located in the middle of the Arabian Desert. Rub'al-Khali alone is more than the entire country of France.

Some examples of animals that reside inside the Arabian Desert include spiny-tailed lizards gazelles, sand cats and oryx. Domestication of camels started around 3500 years ago. It is very likely to be thought that the Arabian Desert was one of the first, if not the first place which this took place.

The Arabian Desert has two major regions. In the east there are sedimentary rocks which have formed over the last 554 million years, they are found in marine basins as well as on continental shelves. To the west you can find The Arabian Platform. It is part of African Shield.

Like all deserts, warm (like that of the Arabian Desert) and cold and cold, the Arabian Desert is very windy. If you live in hot, sandy areas like the desert, this could lead to, and can result in severe storms of sand.

Gobi Desert

Although all deserts are interesting they can be claimed that the Gobi Desert is among the most fascinating. It is located in Asia and located in the Asian continent, Gobi Desert is situated in the Asian continent. Gobi Desert is certainly one of the most unusual deserts in the world. While it's a frigid desert during winter months, where it can get as cold as minus 40°F however, it can be extremely hot during summer and can reach up to temperatures of 122 degrees Fahrenheit. It is likely that you will notice this is distinct from the other deserts that we have reviewed, in that each is mostly hot or cold desert (although it is the case that hot deserts do decrease in temperature during the night).

Gobi Desert Gobi Desert is declared an arid desert (a location with very low rainfall) because the adjacent Himalayas prevent it from rain. It receives only about seven inches of rain per year. In rare instances, during extremely cold winters in the Gobi Desert may see snow. It is not common, but.

It is believed that the Gobi Desert may be the most extensive dry region in Asia yet it is the fifth-largest desert in the world.

Kalahari Desert

Similar to the Sahara Like the Sahara, the Kalahari is one of the world's most renowned deserts. The Kalahair is located in the southern part of Africa and covers huge portions of Botswana as well as portions in Namibia as well as South Africa. The Kalahair is in reality a semi-desert. It includes areas of non-arid land that receive decent rainfall that can sustain the lives of animals and plants more than a complete desert could. Certain areas in the Kalahari are more desert-like like deserts than others. The Kalahari has more diversity in terms of the different climate gradations than typical deserts. Below the Kalahair are vast subterranean reservoirs of water. It is believed that they are remnants of earlier lakes.

The Kalahari therefore has a broad assortment of both plants and animals. One of the most distinctive animals is meerkat. Meerkats are a very small and social animal belonging to the mongoose species.

Patagonian Desert

It is the Patagonian Desert, which is located in Argentina, is biggest desert in Argentina and is also it is the 7th largest desert in the world. It is one of the desert that is a winter (cold) desert and is mostly located in Chile however some parts are also found in Chile. The temperature that is the highest can be around 12 Celsius and the lowest temperature can be about 3 Celsius. Frost is prevalent throughout The Patagonian Desert, however snow isn't, since the desert is extremely dry and receives very low precipitation. Although the Patagonian Desert is a rough environment, there are many animals that reside there, whether at all times or for a portion all the time.

Great Victoria Desert

The Great Victoria Desert is a subtropical desert that is located mostly across South Australia and Western Australia. It is the biggest of the deserts in Australia. It has grassland plains, sandhills that are of smaller dimensions, salt lakes and areas with densely packed pebbles on their top layer. This Great Victoria Desert has very small amounts of mammals and large birds. This desert is able to receive between 7.9 up to 9.8 inches of rainfall per year, but it is home to a significant number of storms, which is unusual for a desert. It experiences about 15 to 20 storms every year. Temperatures in summer are extremely extreme heat, with an average between 32 Celsius to 40 Celsius as well as winter temperatures that vary from 19 Celsius up to 22 Celsius.

Syrian Desert

It is Syrian Desert is a subtropical desert that lies in the southwest of Asia that is expanding to the north, across to the Arabian Peninsula to a large area north of Saudi Arabia, western Iraq and eastern Jordan as well as southern Syria. It is believed that the Syrian Desert receives fewer than five inches of rainfall a year. The surface of the Syrian Desert is extremely smooth and hard and rocky. This desert experiences hot summers and warm winters.

Great Basin Desert

The Great Basin Desert is the largest desert in the United States, and is situated to the east in the Rocky Mountains, in the American southwest. In the Great Basin Desert, average annual rainfall is anywhere from seven to twelve inches each year. It is cold, because it is affected by an effect of rain shadow. Rainshadows are dry land area that is located near an area of mountainous terrain, located in the midst of the mountain free of winds.

Chihuahuan Desert

It is the Chihuahuan Desert is a subtropical desert that is located in the areas of both the United States and Mexico. Within the United States, it occupies vast areas of southwest Texas and small parts in Arizona in addition to New Mexico. It is the Chihuahuan Desert is the second largest desert in North America, and the third-largest one in the Western hemisphere.

It is believed that Chihuahuan Desert Chihuahuan Desert has the most biodiversity, in regards to the animals and plants. However, excessive grazing has resulted in serious degrading. In the Chihuahuan Desert has summer temperatures typically between 35 and 40 Celsius The desert's winters can be cold or cool and may occasionally suffer frosts. The highest temperatures in the desert tend to be found in valleys and other areas with lower elevations.

20 percent in 20 percent of Chihuahuan Desert is grassland, which is comprised of the grasses as well as shrubs. This Chihuahuan Desert has existed for approximately 8000 years, and therefore is one of the oldest deserts.

Great Sandy Desert

The Great Sandy Desert is a subtropical desert that is located in Australia particularly located in Western Australia and Australia's Northern Territory. In some areas, particularly in the northern regions, rainfall is extremely minimal. In other regions however, it can be up to 12 inches each year. In fact, for a desert in The Great Sandy Desert has quite an excessive amount of rainfall. In the majority parts of Great Sandy Desert, thunderstorms can occur for 20-30 days each year. In the northern regions where there is a lot of rain, it can last from 30 to 40 days of this type every year.

The Great Sandy Desert has numerous animals, including goannas, feral camels, dingos, red kangaroos marsupial moles and a variety of species of lizards and birds. A few of the birds that inhabit the Great Sandy Desert are the scarlet-chested parrot, the Alexandra's Parrot as well as the Mulga Parrot.

Karakum Desert

The Karakum Desert is a cold desert that is winter that is situated within Turkmenistan, Central Asia. Around 70% of Turkmenistan is covered by the Karakam Desert. The desert gets only 70-150 millimetres rain per year.

It is located in the Karakum Desert is located to the east of the Caspian Sea, and to the south of the Aral Sea. The area of the desert is expanding because of being that level of the South Aral Sea is falling. Additionally it is the Kara Kum Canal, the second-largest canal for irrigation in the world is located in the Karakum Desert. In addition there is there is the Trans-Caspian Railway runs across the Karakum Desert. Huge amounts of natural gas and oil have been discovered in the desert.

The Bolshoi (Big) Balkan is an area of mountains that is situated in the desert. Within this mountain range, it has been discovered human remains, which were identified as being dating back to time of the Stone Age.

Sonoran Desert

It is the Sonoran Desert is located in regions of two of the United States (Arizona and California) and Mexico (Sonora and Sinaba). It is subtropical desert. Sonoran Desert Sonoran Desert is notable for its abundance of kinds of animal and plant life that inhabit the boundaries. The desert is home to around 60 mammal species and 350 bird species amphibians, 20 species of bird as well as more than 100 reptile species. The Sonoran Desert has more than 2,000 species of plants. One of the most interesting facts concerning Sonoran Desert Sonoran Desert is the fact that it is home to the sole wild Jaguar population throughout the United States. It is believed that the Jaguar is a kind of large cat.

Chapter 9: "Mirages": One Of The Greatest Desert Mysteries

When we think of deserts one of the most intriguing concepts that spring to your mind is the mirage. The capacity of the desert to create mirages has been mentioned in my television shows as well as films, ranging from dramatic dramas to animated films. Imagine a hapless person wandering around the desert, trying to locate water, eventually, seeing a pool of water, and moving towards it but then realizing that after a considerable duration that the water doesn't exist or is moving and further away. This is obviously impossible and implies that it isn't there.

The illusion of a mirage can be that is caused by the effects of climate. Contrary to popular belief they do not just occur in deserts. They can be seen in different locations, too.

But what exactly is an actual mirage? Mirages occur when there is a refraction (bending) in light air at different temperatures. You may have noticed that warmer air is more dense that colder air. Since it is the more dense air cold air is more able to absorb refractive. The result of this is that bending occurs when light is transferred to the hot atmosphere from cooler air. The bending happens upwards, in directions of more dense air, while in the opposite direction to the ground. What is the mechanism by which this illusion can take place?

For the human eye, these rays that bend appear to come from the ground and so the eye perceives on the ground a distorted image from the heavens. It's similar to a reflection one might observe at the top of the lake. This can cause confusion.

Apart from deserts, another one of the places in which mirages are likely to occur can be the roadway. It is because roads absorb a significant amount of heat. In reality, any location in which the ground can absorb an enormous amount of heat could be capable of sustaining mirages. If the ground is heated to a significant degree then the air directly above it is also heated, which creates a sharp rise in temperature of the air that is one of the conditions that need to be present for a mirage to take place.

Another method of explaining the mystery is to explain it through the reality that light has the capability of traveling at different speeds through various temperatures of air. This could alter our normal experience of light. In other words, the eye only sees sunlight's light waves that have passed directly through the air to us. When light passes through cold air before moving into hot air and then hot air, refracted (bending) occurs. Refraction causes light that is coming from the sky to change shape into a U-shape. However, our brains assume that the light simply arrived at us in its normal straight line. Our brains aren't considering this and think it's the ground, or some other thing that is on that ground are the sources of light.

If you've ever witnessed a an eerie, wet glow hovering over the pavement on a scorching humid day, you've witnessed a growing illusion.

Conclusion

To end, I'd like to add that traveling through the Sahara Desert requires a major shift in perspective. The vastness of the landscape that it's difficult to determine the distance. The palette of colors - thousand shades of greige makes it difficult to identify distinct shapes. The intensity of the reflection of sunlight on the sand can trick the eyes to see things that do not exist. Self-reliance, while a great asset for travelers, can be an opportunity for risk: make one error and you'll be gone for eternity. Engage an experienced guide. Have fun exploring the Sahara Desert!

www.ingramcontent.com/pod-product-compliance
Lightning Source LLC
Chambersburg PA
CBHW050236120526
44590CB00016B/2112